Abdelhafid Mimouni

Peptides and Metals: Bioinorganic deciphering

Abdelhafid Mimouni

Peptides and Metals: Bioinorganic deciphering

ScienciaScripts

Imprint

Cover image: www.ingimage.com

This book is a translation from the original published under ISBN 978-3-639-54469-5.

Publisher:
Sciencia Scripts
is a trademark of
Dodo Books Indian Ocean Ltd. and OmniScriptum S.R.L publishing group

120 High Road, East Finchley, London, N2 9ED, United Kingdom
Str. Armeneasca 28/1, office 1, Chisinau MD-2012, Republic of Moldova, Europe
Managing Directors: Ieva Konstantinova, Victoria Ursu
info@omniscriptum.com

Printed at: see last page
ISBN: 978-620-8-38640-5

"Peptides and Metals: Bioinorganic deciphering".

Author: Dr. Abdelhafid Mimouni: Independent researcher in chemistry

Holder of a PhD in Chemistry from the University of Paris XII

(1997) and a Diplôme des Études Approfondies in Bioinorganic

Systems

from the University of Paris XI (93), a Bachelor's and Master's degree

in chemistry (91,

92).

Abstract: This book explores the growing impact of **metal-peptide complexes** in bioinorganic chemistry and their essential role in modern biology and medicine. It examines the structure, synthesis and properties of **peptides**, their interaction with **essential metals** such as copper, zinc and iron, and their application in vital biological processes. Covering topics such as **enzyme regulation**, **cell protection** and **antimicrobial treatments**, the book presents concrete examples of therapeutic applications in **cancer**, **neurodegenerative diseases** and **gene therapy**. It also explores the technical challenges of designing **selective and stable complexes**, while highlighting **recent advances** facilitating research through **artificial intelligence** and **molecular modeling**. This book is a comprehensive resource for understanding the current challenges and future prospects of metal peptides in biomedical applications.

Table of contents

Introduction

Metal peptides represent one of the most fascinating and promising advances in bioinorganic chemistry, biotechnology and modern medicine. Combining the flexibility of **peptides**, which are chains of amino acids capable of interacting specifically with a variety of biological targets, with the unique properties of **essential metals** in the human body, these complexes play a key role in a wide range of biological processes. Their ability to bind metal ions, modulate protein stability and activity, and intervene in vital mechanisms such as **DNA repair**, **cell metabolism** and **immune response**, makes them indispensable players in **cell biology**.

At the interface of chemistry and biology, research into **metal-peptide complexes** has seen spectacular growth in recent decades. This discipline not only enables us to understand **metal interactions in biological systems**, but also to design innovative therapeutic tools. These **complexes** have found particularly promising applications in fields as diverse as **precision medicine**, **cancer** treatment, **gene therapy** and the fight against **multi-drug-resistant infections**. Their potential is further enhanced by recent advances in **modeling technologies**, **artificial intelligence** and **nanotechnology**, which make it possible to streamline the design and predict the efficacy of new metal-peptide complexes before they are synthesized.

The present book aims to explore this emerging field in depth, providing a comprehensive overview of the **theoretical foundations** and **practical applications** of metal-peptide complexes in the biomedical sciences. We will cover the structure and synthesis of peptides, the mechanisms by which they coordinate with metals, their crucial role in biological regulation, and their use in innovative therapeutic applications. In addition, we will analyze the technical challenges encountered in the design of these complexes, the prospects offered by **artificial intelligence** and **computer modeling**, and finally, we will give an overview of **recent discoveries** and their potential impact on future medical treatments.

This book is aimed at researchers, students and professionals in the fields of **bioinorganic chemistry**, **biomedicine**, **pharmacology** and **biotechnology**, as well as anyone interested in innovation in **molecular therapeutics**. The aim is to provide a comprehensive resource that brings together current knowledge, challenges and future prospects for metal-peptide complexes, a subject that will undoubtedly continue to evolve and play a central role in **biomedical innovation**.

Chapter 1: Peptides: Structures, Synthesis and Properties

1.1 Definition of peptides and proteins

Peptides and **proteins** are biological macromolecules essential to many biological processes, but they differ in size and structure. A peptide is a chain of **amino acid residues** linked by **peptide bonds**. Peptides are generally smaller than proteins, ranging from a few amino acids to around 50 residues. Proteins, on the other hand, are larger macromolecules made up of hundreds to thousands of amino acids. A protein may contain one or more polypeptide chains, which are organized in complex ways to perform specific biological functions.

Amino acids are the building blocks of peptides and proteins. There are 20 standard amino acids, differentiated by their side chains (or R groups). These side chains give each amino acid unique chemical properties, influencing the structure and function of peptides and proteins. The **peptide bond** is formed between the amine group (NH_2) of one amino acid and the carboxyl group (COOH) of another, with the removal of a water molecule (condensation).

1.2 Fundamental structures and biological significance

Peptides and proteins have three-dimensional structures that are crucial to their biological function. These structures are organized according to four levels of complexity:

- **Primary structure**: This is the linear sequence of amino acids in the peptide or protein, determined by the genetic information contained in the DNA.
- **Secondary structure**: polypeptide chain segments fold into regular patterns, such as alpha (α) helices and beta (β) sheets, stabilized by **hydrogen bonds**.
- **Tertiary structure**: This is the overall folding of the polypeptide chain in three-dimensional space. This folding is stabilized by non-covalent interactions, such as **hydrophobic forces**, **ionic bonds** and **disulfide bridges**.
- **Quaternary structure**: Some proteins are made up of several polypeptide chains which join together to form a more complex structure (e.g. hemoglobins are made up of four subunits).

The biological importance of peptides and proteins is immeasurable. They play a crucial role in virtually every biological process: regulation of metabolism, transmission of cellular signals, immune response, catalysis of biochemical reactions (enzymes), and much more besides.

Peptides such as **insulin,** for example, are essential for blood sugar regulation and energy metabolism homeostasis.

1.3 Examples: Insulin and its Key Role in Glucose Metabolism

Insulin is a peptide composed of 51 amino acids, and is one of the best-known examples of a peptide with a vital biological function. It is produced by the **beta cells** of the islets of Langerhans in the pancreas and plays a central role in glucose metabolism. When an individual consumes carbohydrates, the concentration of glucose in the blood rises. In response, insulin is secreted into the bloodstream, where it facilitates the uptake of glucose by cells, notably muscle and liver cells, for use as an energy source or for storage as glycogen.

Insulin also helps regulate the production of **glucagon**, another pancreatic hormone, to maintain a stable concentration of glucose in the blood. Dysfunctions in insulin production or action are at the heart of pathologies such as **type 1 diabetes** (where insulin production is insufficient) and **type 2 diabetes** (where cells become resistant to insulin). This hormone is therefore a key example of a peptide with major implications for biology and medicine.

1.4 Peptide synthesis methods

Peptide synthesis is a procedure for creating artificial peptides in the laboratory. There are a number of methods available to achieve this synthesis, depending on the researcher's needs and the peptide's characteristics.

- **Solid-resin peptide synthesis (SPPS)**: **Solid-resin peptide synthesis (SPPS)** is the most commonly used method for producing peptides in the laboratory. It is based on sequential construction of the peptide chain on a solid resin, with successive addition of protected amino acid residues. Each amino acid is added one by one, and after each reaction step, the chain is washed to eliminate by-products. Once the sequence is complete, the peptide is cleaved from the resin and purified. This method is rapid, but may have limitations in the case of long or difficult-to-synthesize peptides.
- **Chemical synthesis**: There are also **chemical methods** that use reagents to bind amino acids in solution. These methods can be adapted for specific, complex peptides, particularly where chemical modifications are required. They also allow

modification of the peptide chain, such as the introduction of non-natural residues or post-translational modifications.

- **Expression in biological systems**: For more complex peptides, or when large quantities are required, **recombinant synthesis** via biological systems (bacteria, yeast or mammalian cells) can be used. In this process, the gene encoding the peptide of interest is inserted into a vector, which is then introduced into a host cell. The host cell then produces the peptide, which can then be purified. This method is particularly useful for large peptides, or when a modified or specific form is required.

1.5 Advances in Peptide Modification for Biomedical Applications

Peptides have great potential for **biomedical applications**. However, to make them more effective, stable and specific, several **chemical** and **structural modifications** are often introduced into their design.

- **Peptides modified for stability**: Peptides can be rapidly degraded by enzymes in the body. To improve their **stability**, modifications such as the incorporation **of analogues of the peptide bond** (e.g. amides, or the use of D-amino acids) are often carried out. This increases their shelf life in the body and enhances their therapeutic efficacy.

- **Modification for specificity**: Peptides can also be modified to increase their **specificity** towards specific biological targets, such as membrane receptors, proteins or biomolecules involved in pathological processes. For example, **immunomodulatory** or **antimicrobial** peptides have been developed to treat infectious diseases or immune system disorders.
- **Therapeutic peptides and biomaterials**: Peptides are also used in the design of **biomaterials** and **nanomedicine**. Their ability to bind to metals, self-assemble or interact with specific biological systems makes them particularly interesting for **drug delivery**, medical imaging and biomedical devices.

Chapter 2: Metals in Biology

2.1 Essential metals in biological systems

Metals play a fundamental role in biological systems. They are involved in numerous biochemical functions, from enzyme catalysis to the regulation of gene expression. These metallic elements, often referred to as **trace** or **essential metals**, are necessary for life, albeit often in small quantities. Essential metals in biological systems include :

- **Copper (Cu)**: Copper is an essential transition metal in many enzymes and proteins. It is involved in processes such as **cellular respiration**, **connective tissue formation** and **free radical detoxification**.
- **Iron (Fe)**: Iron is one of the most studied metals in biology, due to its critical role in oxygen transport and **redox** reactions. It is essential for proteins such as **hemoglobin** and **myoglobin**.
- **Zinc (Zn)**: Although often present in small quantities, zinc plays a fundamental role in the structure and function of many **metal proteins**. It is involved in **enzyme catalysis**, regulation of **gene expression** and stabilization of protein structures.

- **Cobalt (Co)**: Cobalt is a key element in **cobalamin** (vitamin B12), a coenzyme essential for **nucleic acid** synthesis and energy metabolism.
- **Actinides and lanthanides**: Although not as widely studied in biological systems as the previously mentioned metals, some actinides and lanthanides (such as **uranium**, **thorium**, **cerium** and **lanthanum**) can interact with biological systems in very specific contexts, although their biological role is less established.

The importance of these metals lies not only in their presence, but also in their ability to interact with biological ligands (such as peptides and proteins) to form **metal complexes**, which are essential to the regulation and activity of many biological processes.

2.2 Examples: Iron in Hemoglobin and Copper in Cytochrome C Oxidase

Metals are often at the heart of complex biochemical structures that perform vital biological functions. Two emblematic examples are :

- **Iron in hemoglobin**: Hemoglobin, the protein responsible for oxygen transport in the blood, contains **iron** atoms **(Fe^{2+})** in a heme group, a structure that binds reversibly to oxygen. This ability of iron to bind oxygen and release it into tissues is crucial to **respiratory gas homeostasis**.

- **Copper in cytochrome C oxidase**: Copper plays an essential role in the **mitochondrial electron transport chain**. **Cytochrome C oxidase**, an enzyme complex containing copper atoms, is responsible for reducing **molecular oxygen** to **water**, a key step in ATP production via **oxidative phosphorylation**.

These two examples illustrate the importance of iron and copper in vital processes such as cellular respiration and respiratory gas transport, highlighting how metals act as catalytic elements in complex biological systems.

2.3 Chemical properties of metals and their role in cellular biochemistry

Essential metals have unique chemical properties that make them suitable for specific biological functions. These properties include their ability to form **metal complexes** with organic ligands (proteins, peptides, DNA), as well as to participate in **redox reactions**. Here are a few important properties of metals in biology:

- **Redox properties**: Iron, copper and other transition metals are particularly well known for their ability to undergo changes in oxidation state. These changes are fundamental to the catalysis of biochemical reactions, as in **oxidoreductase** enzymes (e.g. **cytochrome C oxidase** and **reductase**).
- **Coordination and complexation**: Transition metals have the ability to **coordinate** ligands, i.e. to form complexes where the metal is bound to nitrogen, oxygen or sulfur atoms of ligands (from amino acids, peptides or nucleic acids). For example, **zinc** in **metalloproteins** (such as **zincoenzymes**) plays an essential role in enzyme catalysis, stabilizing charges and facilitating biochemical reactions.
- **Acid-base behavior**: Certain metals, notably zinc and copper, are involved in regulating pH in biological systems, which is essential for maintaining the acid-base balance of cells and tissues.

These properties enable metals to perform a variety of crucial functions in **cellular biochemistry**. For example, metalloprotein enzymes, which contain metals in their active site, catalyze biochemical reactions such as hydrolysis, reduction and oxidation, facilitating processes such as protein synthesis, nutrient degradation and metabolic regulation.

2.4 Zinc in metalloproteins and silver in antimicrobial peptides

Zinc in metalloproteins: Zinc is an essential metal in a variety of **metalloproteins**. It is often present in enzymes such as **carbonic anhydrases** and **peptidases**. Zinc plays a stabilizing role in protein structure and is involved in **enzyme catalysis**. For example, in **DNA repair enzymes**, zinc is used to facilitate the repair of DNA strand breaks, which is crucial for the maintenance of genetic integrity.

Silver in antimicrobial peptides: **Silver** is a metal with notable antimicrobial properties. **Antimicrobial peptides** containing silver are used in a variety of medical applications to combat bacterial infections. Silver exerts a biocidal action by disrupting bacterial cell membranes and interfering with their enzymatic processes. For example, silver-peptide complexes are increasingly used in **antimicrobial dressings** and **medical devices** to prevent nosocomial infections.

These examples demonstrate the importance of metals not only in basic enzymatic and biological functions, but also in innovative biomedical applications, such as infection control and the management of diseases linked to metal deficiencies.

Chapter 3: Metal-Peptide Complex Formation

3.1 Peptide Coordination Mechanisms with Metals: Monodentate, Bidentate, Polydentate

The formation of **metal-peptide complexes** relies on the ability of peptides to interact with metal ions via **coordination** bonds. Peptides, which are made up of amino acid chains, can bind to metals via functional groups on their residues, such as amine, carboxyl, sulfur (thiol) or nitrogen groups. This interaction, known as **coordination**, can occur in different ways, depending on the number of coordination sites available in the peptide chain and the nature of the metal involved.

- **Monodentate coordination**: In this case, only one atom or functional group of a peptide binds to the metal ion. For example, an amine group ($-NH_2$) can bind to a metal ion, forming a **monodentate complex**. This type of binding is generally weaker and less stable, but can be useful in systems where a temporary or reversible interaction is required.
- **Bidentate coordination**: Here, two atoms or functional groups of the peptide bond to a central metal. This forms a **bidentate complex**, which is more stable than the monodentate complex due to **chelation**. For example, amino acid residues such as

histidine can bind to the metal via their nitrogen and sulfur or carboxyl atom, enabling the formation of bidentate complexes.

- **Polydentate coordination**: This type of coordination involves several atoms or functional groups of the peptide bonding to a metal ion, forming a **polydentate complex**. Peptides with residues such as amino acids containing thiol groups (like **cysteine**) or carboxylate groups can interact more complexly and stably with the metal. This type of coordination enables very strong interactions and is crucial for the formation of metal complexes in many biological functions, notably in metalloprotein enzymes.

The peptides that can form these types of complexes are called **ligands**, and play an essential role in **stabilizing metals** in biological systems. The diversity of these interactions (monodentate, bidentate and polydentate) enables high **specificity** and **selectivity**, tailored to the biochemical needs of the organism.

3.2 Examples: Copper Complex with Cyclodextrin in Cancer Therapy

One of the most interesting examples of metal-peptide complexes in biology is **the copper complex with cyclodextrin**. **Cyclodextrin** is a small cyclic carbohydrate structure with unique metal **chelation** properties. This property is being exploited in a number of therapeutic strategies, notably in **cancer therapy**.

- **Copper (Cu^{2+}) is** an essential metal involved in many biological processes, including angiogenesis (the formation of new blood vessels), a key process in the progression of cancerous tumors. Excess copper in tumor cells promotes tumor growth and resistance to anti-cancer treatments.
- In this context, **cyclodextrin** is used to form a **copper-cyclodextrin complex**, which enables copper to be transported in a controlled manner to cancer cells. This complex is capable of slowly releasing copper into targeted cells, thereby inhibiting tumor cell growth while minimizing toxic effects on normal cells.

This mechanism of **targeted copper delivery** has opened up new avenues in cancer treatment, exploiting the ability of metal-peptide complexes to deliver metals to specific areas of the body and interact with biological targets in precise ways.

3.3 Role of Peptides in Metal Ion Stabilization: Thermodynamic Stability and Specificity

Peptides play a crucial role in **stabilizing metal ions** in various biological processes. This ability is determined by several factors, including the **peptide's specificity** for certain metals and the thermodynamic stability of the complexes they form.

- **Thermodynamic stability**: A metal-peptide complex must be sufficiently stable to maintain the metal ion in solution and avoid precipitation or loss through degradation. The **thermodynamic stability** of metal complexes depends on **electrostatic interactions**, **chelation** (formation of multi-site complexes), and the **availability of** ligands in the biological environment.
- **Peptide specificity**: Peptide **specificity** for certain metals is another key aspect of metal ion stabilization. For example, peptides such as **metalloproteins** contain residues that are specifically adapted to bind metals such as zinc, copper or iron, forming extremely stable complexes. This specificity is crucial for biological functions such as metal transport, storage or use in enzymatic reactions.

Peptides also regulate **the release** and **uptake** of metals, ensuring that metal ions do not accumulate to toxic concentrations in cells. For example, proteins such as **ferritin** and **transferrin** are metalloproteins that stabilize metal ions (such as iron) in the body by storing them or facilitating their transport.

3.4 Application of Peptides in Metal Storage Capsules in Biological Processes

Another important area of application for metal peptides concerns their use in the **storage and management of metals** within biological systems. Many proteins and peptides act as **storage capsules** for metal ions, playing a key role in the control of metals in the organism.

- **Ferritin**: Ferritin is a protein that plays an essential role in the **storage of iron** in cells. It stores iron in a solubilized, bioavailable form, reducing the risk of toxicity associated with excessive free iron. Iron is stored in a **ferric core** within ferritin, and can be released when required for biological processes such as **hemoglobin synthesis**.
- **Metallothioneins**: These are **cysteine-rich** peptides that can bind metal ions such as zinc, copper and cadmium. They play a key role in the **detoxification of** heavy metals and the regulation

of zinc and copper metabolism, and are involved in **cellular protection** mechanisms against oxidative stress.

- **Biomolecular capsules**: Peptides can also be used in **biomolecular capsules for the controlled storage of** metals in **nanomedicine** and **drug delivery** applications. For example, modified peptides can be used to encapsulate metal ions in nanoscale structures and deliver them in a targeted manner to specific tissues.

These mechanisms show how peptides play a key role in controlling **metal availability** in biological systems, a crucial factor in many biological and therapeutic processes.

3.5 Dynamics of Metal-Peptide Complex Formation

The **dynamic formation** of metal-peptide complexes is a crucial process in biomolecular interactions. Peptides can bind to a metal ion in a reversible manner, allowing affinity and specificity to be modulated according to environmental conditions. Factors influencing these dynamics include **pH**, **metal concentration**, and the **presence of other ligands** that may interfere with coordination sites.

Dynamic chelation mechanisms are particularly important in biological systems, as they allow **fine control of metal availability**, essential for processes such as metal transport into cells, regulation of the immune response, or detoxification. For example, the **stability** of copper complexes with peptides can be influenced by changes in pH, enabling the system to tailor copper binding to biological needs.

3.6 Importance of Metal-Peptide Complexes in Bioinorganic Catalysis

Another key application of metal-peptide complexes is their role in **bioinorganic catalysis**. Peptide complexes with transition metals can act as **biomimetic catalysts**, accelerating specific biochemical reactions, such as the **oxidation** or **reduction** of certain molecules. This capability is exploited in processes such as pollutant degradation, drug synthesis or catalysis in biological systems.

Thanks to their flexible structures and specific coordination sites, peptides can induce conformational changes that favor the activation of metal catalytic sites. For example, metal-peptide complexes containing

copper or **zinc** can catalyze **reduction** or **oxidation** reactions, playing a role similar to that of metalloprotein enzymes in biological systems.

These complexes can also be designed to operate under **mild conditions**, making them suitable for applications in green catalysis or in sensitive biological environments, such as living cells or human tissue.

3.7 Applications of Metal-Peptide Complexes in Medicine

Metal-peptide complexes have considerable potential in the field of **medicine**, particularly in **targeted therapies** and **cancer treatments**. Because of their ability to bind specifically to metal ions, peptides can be used to create **targeted drugs** that deliver metal ions to specific cells or tissues, offering more precise and effective treatment.

An example of this application is the use of metal-peptide complexes in **cancer therapy**. Metal peptides can be designed to bind to specific tumor cells, then release metal ions that interact with proteins essential for tumor growth. **Copper** and **platinum**, for example, are metals widely used in anti-cancer treatments. In this context, copper or platinum complexes associated with peptides can bind to specific sites on cancer cells and release metals, inducing cytotoxic effects.

Metal-peptide complexes are also being studied for their potential in **antimicrobial** treatments, particularly in the fight against antibiotic-resistant pathogens. Antimicrobial peptides that bind to metal ions can disrupt the cell wall of microbes or interfere with essential biological processes, offering a new approach to treating hard-to-eradicate infections.

3.8 Industrial applications of Metal-Peptide Complexes

Apart from biology and medicine, metal-peptide complexes are also finding applications in industrial sectors such as **biotechnology**, **energy** and **environmental protection**. Their ability to capture, store and release specific metals in a controlled manner can be used in processes such as **metal resource management**, **biomethallurgy** and **wastewater treatment**.

For example, certain metal-peptide complexes can be used in **carbon capture systems**, where metals such as **copper** or **nickel** play a role in converting carbon dioxide (CO_2) into organic molecules, in a process similar to photosynthesis. In biomethallurgy, metal peptides can be employed to extract precious or rare metals more selectively and efficiently, reducing the environmental impact of traditional mining.

Chapter 4: Metal Peptides in Biological Regulation

Metal peptides play a key role in the **biological regulation** of cellular processes, in particular as modulators of enzyme activity and players in complex cell-signaling mechanisms. These complexes, which result from the interaction of peptides with metal ions, are involved in a wide range of physiological functions, from the regulation of enzyme activity to intracellular signalling and the response to oxidative stress.

4.1 Modulation of Enzyme Activity by Metal-Peptide Complexes

Metal-peptide complexes are essential for modulating enzyme activity, particularly in processes where **transition metals** play a catalytic role. Metal ions, when coordinated to peptides, can activate or inhibit certain enzymes depending on their ability to interact with specific active sites. For example, **zinc**, **copper** and **manganese** are often involved in the regulation of **metalloproteins** and **metal enzymes**.

A key example is the inhibition of **metal proteases** and **phosphatases** by metal peptides. The latter can bind to metal ions present in the enzymes' active sites and modify their activity, inhibiting or activating their function. For example, certain peptides containing **zinc** or **copper** ligands are used to inhibit specific enzymes involved in pathological processes, such as neurodegenerative diseases.

4.2 Peptidomimetics and their interaction with metals

Peptidomimetics are synthetic molecules that mimic the structure and function of natural peptides, while offering greater stability and ease of modification. These molecules can bind to specific metals to induce targeted therapeutic effects. In the context of **neurodegenerative diseases** such as **Alzheimer's** and **Parkinson's**, peptidomimetics have been developed to interact with metals such as **copper** or **zinc** and modulate the activity of enzymes involved in the pathology of these diseases.

A well-known example is the use of peptidomimetics containing coordination sites for **copper** to inhibit the enzyme **beta-secretase**, involved in the formation of amyloid plaques in the brains of Alzheimer's patients. Inhibition of this enzyme by metal peptides has the potential to slow disease progression by reducing plaque formation and alleviating neurodegenerative symptoms.

4.3 Metal Peptides in Cell Signaling Mechanisms

Metal peptides also play a fundamental role in **cell signalling**, a process essential to cell-to-cell communication and the regulation of complex biological functions. Metal ions, such as **calcium, zinc** or **copper**, act as **second messengers** in several intracellular signaling pathways, influencing processes such as **cell proliferation, differentiation** and **stress response**.

Zinc-peptide complexes, in particular, are involved in the regulation of **intracellular receptors** and **DNA repair**. For example, peptides containing zinc ions can interact with **nuclear receptors** and modulate their activity in **DNA repair** after free radical or radiation damage. This interaction between zinc and peptides could promote **DNA damage repair** by activating proteins involved in **base excision repair** or **homologous recombination repair**, both essential mechanisms for maintaining genomic integrity.

4.4 Recent discovery : The Use of the Zinc-Peptide Complex in DNA Repair

Recent research has highlighted the use of **zinc-peptide complexes** in DNA repair. **Zinc**, as an essential metal in several enzymes involved in DNA repair, can bind to peptides to activate or stabilize repair proteins. For example, some zinc-containing peptides bind to **DNA repair**

receptors, such as **proteins of the zinc-finger family**, to induce repair of DNA strand breaks. These complexes can enhance **DNA lesion recognition** and facilitate **damage repair**, which is crucial for maintaining genomic stability and preventing mutation-related diseases such as cancer.

An interesting therapeutic application of these zinc-peptide complexes could be their use in **the treatment of cancer**, where rapid repair of tumor cell DNA is essential for cell survival. By targeting the DNA repair mechanisms of cancer cells, it would be possible to increase their susceptibility to chemotherapeutic treatments or radiation, while reducing the damage caused to normal cells.

Chapter 5: Metal-Peptide Complexes and Antimicrobial Properties

Metal-peptide complexes represent a promising approach to combating **bacterial** infections, particularly in the face of rising antimicrobial resistance. These complexes exploit the unique properties of peptides, which can interact with metal ions to create structures capable of destabilizing bacterial cell membranes and inhibiting their growth. **Copper,** in particular, is an essential metal in this application, due to its biocidal properties and its ability to interact effectively with antimicrobial peptides to enhance their activity.

5.1 Antimicrobial peptides and Metal-Peptide Interactions : The Role of Copper-Peptide Complexes in Bacterial Infection Control

Antimicrobial peptides (AMPs) are small chains of amino acids that possess antimicrobial activity against a wide range of pathogens, including bacteria, viruses and fungi. These peptides exert their activity through a variety of mechanisms, including disruption of cell membranes, inhibition of protein synthesis, or alteration of microbial cellular metabolism.

When an antimicrobial peptide binds to a metal ion, such as **copper (Cu^{2+}), the formation of metal-peptide complexes** can significantly increase the peptide's efficacy. Copper, in particular, possesses notable anti-infective properties due to its ability to generate **free radicals** via redox reactions, disrupting membrane integrity and inhibiting enzymatic functions essential to bacteria.

Copper-peptide complexes exploit this ability by directing metal ions to specific sites on the bacterial cell membrane, enabling the peptide to better integrate into the membrane and induce its degradation. In addition, copper ions can bind to specific sites on antimicrobial peptides, enhancing their antimicrobial power while reducing the risk of bacterial resistance.

5.2 Concrete examples: The Polymyxin B Peptide and its Complexes with Copper to Improve Efficacy Against Resistant Bacteria

Polymyxin B is an antimicrobial peptide well known for its activity against gram-negative bacteria, including multi-resistant pathogens such as **Pseudomonas aeruginosa** and **Escherichia coli**. This peptide acts primarily by interacting with bacterial membrane **phospholipids**, leading to loss of membrane integrity and cell death.

Polymyxin B complexes with Cu^{2+} ions have shown promising results in **amplifying antimicrobial activity** against resistant bacterial strains. Copper binding to the peptide promotes **integration of the peptide into the lipid membrane**, overcoming some forms of antimicrobial resistance. In addition, the copper ion induces the production of **hydroxyl radicals** and other reactive oxygen species (ROS), causing further damage to bacterial cells and contributing to overall treatment efficacy. This phenomenon is particularly important in the fight against bacteria such as **Klebsiella pneumoniae** or **Acinetobacter baumannii**, which are frequently responsible for serious hospital-acquired infections.

5.3 Studies on Antimicrobial Peptides and their Ability to Penetrate Cell Membranes

A key feature of antimicrobial peptides is their ability to cross **cell membranes**, enabling them to reach their intracellular targets. **Metal-peptide complexes**, particularly those containing **copper** ions, enhance this penetration ability due to the metal's interaction with the lipid membrane, facilitating the peptide's entry into the target cell.

Studies on the **penetration mechanisms** of metal-peptide complexes have shown that the metal ion enhances the peptide's interaction with the **lipid bilayer** of bacterial membranes, promoting the formation of **pores** or **channels** in the membrane. Once inside, metal-peptide complexes can alter the cell's internal functions, for example by interfering with **protein synthesis**, **DNA** or the bacterium's **energy metabolism**, leading to its death.

Research has shown that the effectiveness of copper-peptide complexes lies in their ability to generate **free radicals** that increase cell membrane permeability, enabling antimicrobial peptides to reach their target more easily. These complexes are therefore able to penetrate bacterial membranes effectively, while minimizing potential side effects on human cells.

5.4 Landmark discovery: The Series of lacticin-derived peptides for their antimicrobial efficacy when complexed with Cu^{2+} ions.

A landmark discovery in the field of antimicrobial peptides complexed with metal ions is the use of peptides derived from **lacticin**, an antimicrobial peptide **of lactobacillary** origin. When combined with **Cu^{2+}** ions, these peptides have shown **enhanced antimicrobial activity**, particularly against bacterial strains resistant to conventional antibiotics.

Lacticins, which are cationic peptides, can bind to **copper** ions to form stable complexes that enhance their ability to interact with bacterial membranes. Once the complex is formed, copper ions play a catalytic role in the generation of **free radicals** and other reactive oxygen species, which will disrupt membranes and lead to the collapse of the bacterial cell. Furthermore, lacticin peptides complexed with copper are able to penetrate cell membranes more easily and destabilize critical bacterial biological functions.

This approach could provide a new therapeutic avenue for treating infections caused by multi-resistant bacteria, particularly in the context of **antibiotic resistance**.

Chapter 6: Applications in Medicine and Genomic Therapy

Peptides and **metal-peptide complexes** are playing an increasingly central role in **medicine**, particularly in **cancer treatment** and **gene therapy**. Their ability to interact specifically with biological targets and to be easily modified for targeted action makes them invaluable for the development of new therapeutic approaches. In this chapter, we explore how peptide chemistry, combined with metal ions, opens the way to innovative treatments and more effective gene therapy strategies.

6.1 Peptide chemistry in medicine: Therapeutic peptides and their use in cancer treatment

Therapeutic peptides are increasingly used in the treatment of various pathologies, including cancer. Thanks to their small size, flexibility and ability to bind specifically to biological targets, peptides are able to modulate biological pathways with precision. Unlike traditional small molecules, peptides can target specific receptors or proteins, minimizing unwanted side effects while maximizing treatment efficacy.

In cancer, therapeutic peptides can be used for several purposes:

- **Inhibit tumor growth** by targeting specific receptors or proteins that promote cell proliferation.
- Targeted **induction of cell death** (apoptosis).
- **Improve drug delivery** by increasing penetration into cancer cells.
- **Help repair tissue** damaged by chemotherapy.

Examples of therapeutic peptides include **antitumor** peptides and **peptidomimetics** that act by inhibiting oncogenic proteins, such as **Bcl-2**, which are often overexpressed in tumor cells.

6.2 Application of Metal-Peptide Complexes in Cancer Therapy: Platinum-Peptide Complexes in Chemotherapy and their Mechanism of Action

Metal-peptide complexes have shown promising results in the field of cancer chemotherapy, particularly **platinum-peptide complexes**. **Platinum**, as a transition metal, has been used for decades in chemotherapeutic agents such as **cisplatin**, which is known for its ability to interact with DNA and inhibit cell replication. However, cisplatin has significant side effects, including renal toxicity.

Combining **peptides with platinum ions** overcomes some of these drawbacks. Indeed, the peptide can help to **specifically target tumor cells**, thus improving the efficacy of chemotherapy while reducing side effects. The mechanism of action of platinum-peptide complexes is based on several steps:

1. **Binding to DNA complex**: Platinum binds to tumor cell DNA, forming covalent bridges that block DNA replication and transcription, leading to cell death by apoptosis.
2. **Specific targeting**: Thanks to its ability to bind to specific receptors or sites on cancer cells, the peptide enables platinum to be delivered directly to tumor cells, increasing treatment **selectivity**.
3. **Enhanced cell penetration**: certain peptides can facilitate the entry of platinum-peptide complexes into tumor cells, thereby improving their efficacy.

Recent studies have shown that these complexes can be used to treat a range of cancers, including **breast**, **lung** and **ovarian**. For example, the **cysteine-based platinum-peptide complex** has been explored for its ability to specifically target cisplatin-resistant **cancers**.

6.3 Peptides and Gene Therapy: Using Metal Peptides for Targeted Gene Delivery

Gene therapy represents a revolutionary approach to the treatment of genetic diseases and certain types of cancer. The idea is to insert, modify or delete genes within an individual's cells to treat or prevent disease. However, one of the main challenges of gene therapy is to find an effective and safe way of delivering the genes into the target cells.

Metal peptides have emerged as potential vectors **for targeted gene delivery**. These peptides, when associated with metal ions, can interact with **nanoparticles** or **gene vectors** to facilitate their transport and penetration into cells.

Metal-peptide complexes can be used for :

- **Targeting specific receptors** on the surface of tumor cells or other diseased cells.
- **Protect therapeutic genes** as they are transported through the body, allowing them to cross biological barriers such as cell membranes and the blood-brain barrier.
- **Facilitate the integration of genetic material** into the target cell's genome.

For example, **metal peptide complexes**, such as those containing **copper** or **zinc**, are used to deliver **plasmid DNA** into cancer cells to induce expression of genes that promote **DNA repair** or enhance the immune response against the tumor.

6.4 Concrete example: Copper Peptide Nanoparticles for Genetic Drug Delivery in Cancer Cells

A particularly innovative application of metal-peptide complexes in medicine is the use of **copper peptide nanoparticles** for **genetic drug** delivery. These nanoparticles can encapsulate **therapeutic DNA**, such as genes coding for anti-tumor proteins, and deliver them directly into cancer cells.

Copper-peptide nanoparticles are particularly interesting for **cancer gene therapy** because they can :

- **Selectively target tumor cells** through peptide adhesion to specific receptors on the cell surface.
- **Improve intracellular penetration of** therapeutic genes, by facilitating their entry into cells via endocytosis or membrane fusion.

- **Induce a** local **biological response**, thanks to copper's pro-oxidant effect, which can damage the DNA of tumor cells, thus enhancing treatment efficacy.

Recent studies have demonstrated that copper-peptide nanoparticles can be used to deliver **gene vectors** into cancer cells, resulting in targeted expression of antitumor genes and promoting selective destruction of cancer cells while reducing side effects on healthy cells.

Chapter 7: Peptides and Metals in Cellular Protection

Peptides and **metals** play an essential role in protecting cells against various biological stresses, notably **oxidative stress**. This phenomenon results from an imbalance between the production of **free radicals** and the body's defense capacity. Prolonged exposure to oxidative stress can lead to damage to DNA, proteins and lipids, which in turn can promote the onset of degenerative diseases, premature aging and certain cancers. **Metal antioxidants**, notably metal-peptide complexes, are key mechanisms in the management of this stress, contributing to **DNA repair** and **cellular protection**.

7.1 Metallic antioxidants: Peptides and metals against oxidative stress

Metal antioxidants are molecules capable of neutralizing **free radicals** and preventing oxidative damage. Peptides, as specific ligands for various **metal ions**, play a crucial role in regulating oxidative stress. The interaction between a peptide and a metal often leads to the formation of complexes capable of trapping and breaking down reactive oxygen species (ROS), thereby minimizing cellular damage.

Certain transition metals, such as **zinc (Zn^{2+})**, **copper (Cu^{2+})** and **manganese (Mn^{2+})**, are particularly effective in this role, as they are involved in **antioxidant enzymes**, such as **superoxide dismutases (SODs)**, which catalyze the dismutation of superoxide radicals into less reactive molecules, such as **hydrogen peroxide**. Peptides that bind to these metals help maintain the stability of these enzymatic complexes, enhancing their activity and offering **cellular protection** against free radicals.

7.2 Concrete examples : The Role of Metal-Peptide Complexes in Free Radical Protection

One of the most emblematic examples of cellular protection via metal-peptide complexes is **superoxide dismutase (SOD)**, an enzyme involved in the management of oxidative stress. **SOD** is found in various metal forms, including **Cu/Zn-SOD** (containing copper and zinc), which plays an essential role in human cells to catalyze the conversion of superoxide radicals (O_2-) into hydrogen peroxide (H_2O_2). The latter can then be degraded by other antioxidant mechanisms.

Metal-peptide complexes facilitate the stability and activity of enzymes like **Cu/Zn-SOD**, helping to maintain an **optimal geometric arrangement** for interaction with free radicals. For example, peptides containing cysteine, histidine or glutamine residues can efficiently bind metal ions and help maintain the enzyme in a catalytically active configuration. These complexes are able to **neutralize free radicals** and prevent oxidative damage in cells, thus protecting essential biological structures such as DNA, proteins and cell membranes.

7.3 Discovery: The Glutathione Peptide and its Complexes with Manganese in Cellular Protection Against Oxidative Stress

A recent discovery in the field of **metal antioxidants** has focused on the peptide **glutathione**, a tripeptide composed of **glutamate**, **cysteine** and **glycine**, which plays a major role in cellular defense against oxidative stress. This peptide can bind to metal ions, notably **manganese (Mn^{2+})**, to form stable metal complexes that enhance the effectiveness of the antioxidant response.

Glutathione-manganese complexes have shown promising results in **protecting against free radicals** by modulating the activity of antioxidant enzymes and activating oxidative stress detoxification pathways. These complexes help **minimize oxidative damage** in cells, particularly under stressful conditions, and may be involved in pathophysiological processes such as **neurodegeneration**, where oxidative stress is a major factor in the progression of diseases such as **Parkinson's**.

7.4 Metals in DNA Repair: Examples of Cu/Zn-SOD and its Complexes in DNA Repair and Oxidative Stress Management

In addition to their role in protecting cells against oxidative stress, certain metals, particularly **copper** and **zinc**, also play an essential role in **DNA repair**. **Antioxidant enzymology** is intrinsically linked to DNA repair, as free radicals generated by oxidative stress can cause DNA damage. **Cu/Zn-SOD** is an excellent example of a metal-peptide complex involved in DNA repair.

Copper and zinc, through their presence in **SOD**, are essential for the elimination of superoxides, which, if left unchecked, can damage DNA, leading to mutations and genetic abnormalities. These metals, combined with peptides, help regulate the activation of DNA repair pathways, such as **base excision repair** or **homologous recombination repair**. By maintaining DNA integrity and minimizing cellular damage, these complexes play a crucial role in preventing **genetic mutations** and DNA-related diseases such as **cancer**.

Chapter 8: Technical Challenges in the Design of Metal-Peptide Complexes

Metal-peptide complexes offer immense potential in biomedical applications, including cancer treatment, infection control and regulation of biological processes. However, the design of these complexes presents significant technical challenges, particularly with regard to **stability**, **selectivity**, and their behavior in **complex biological environments**. This chapter explores these challenges and proposes possible solutions by combining **experimental** and **computational approaches** to streamline the design of efficient metal-peptide complexes.

8.1 Stability and Selectivity of Metal-Peptide Complexes

The stability of metal-peptide complexes is a key factor in their effectiveness in biological applications. Metal-peptide complexes must be sufficiently stable to withstand biological conditions, such as pH variations, the presence of competitive ions, and interactions with other biomolecules. **Instability** could lead to dissociation of the metal, degradation of the peptide or formation of by-products that are not biologically active or could cause undesirable side-effects.

In addition, the **selectivity** of metal-peptide complexes is another major challenge. In the biological context, several types of metal ions are present, and these ions can bind to peptides in non-specific ways, potentially leading to **competition** between metals for coordination sites. For example, copper, zinc and iron are often present in biological systems at high concentrations and can interact unpredictably with peptides. It is therefore crucial to design metal-peptide complexes that are not only stable but also **selective**, binding preferentially to a particular metal ion.

8.2 The challenges of designing selective metal peptides in complex biological environments

Biological environments present unique challenges for the design of metal-peptide complexes, including the presence of **variability** in biochemical conditions and competing molecules. The biological system is a dynamic and complex environment, with varying metal concentrations, competing ions, fluctuating pH and interactions with other macromolecules such as **proteins**, **lipids** or **DNA**. These factors can make it difficult to predict the behavior of metal-peptide complexes under real-life conditions.

Physiological properties such as **peptide transport** across cell membranes and **intracellular distribution** are also important considerations. For example, certain metal-peptide complexes must be able to bind to specific cell receptors or cross **biological barriers** such as cell membranes or the **blood-brain barrier** to deliver drugs or genes in a targeted manner. This requires precise design of the peptide, its metal interactions and its physico-chemical characteristics.

A further difficulty lies in the **specificity of the metal-peptide bond**, which can be influenced by factors such as the **functional groups present in peptides** (e.g. cysteine, histidine or glutamine residues). These residues can bind to different metal ions and, depending on the peptide's conformation, these complexes can become either highly specific or non-specific. It is therefore crucial to control **conformation** and **interaction** parameters to ensure specific and stable binding to a particular metal in a complex biological environment.

8.3 Structural modifications and rationalization approaches: An integrated approach to peptide chemistry, computer modeling and in vitro experiments

To overcome the challenges of stability, selectivity, and adaptation to complex biological environments, several strategies have been implemented in the design of metal-peptide complexes. One of the most promising approaches is to combine **targeted structural modifications**, **computational modeling** and **in vitro experiments**.

- **Structural modifications**: Introducing specific chemical modifications into peptides, such as adding new **functional groups** or substituting certain **amino acid residues**, can **direct the coordination** of metal ions and **stabilize the conformation** of the metal-peptide complex. For example, the introduction of histidine or cysteine residues can increase the **specificity of binding** with certain metal ions, while improving the stability of the complex under biological conditions. Techniques such as **peptidomimetics** and the **introduction of chelating groups** can also be used to design more selective complexes.
- **Computational modeling**: Molecular modeling and **molecular dynamics** approaches offer a powerful means of predicting and rationalizing interactions between peptides and metal ions. For example, **docking modeling** can predict the most favorable binding sites for metals, and simulate interactions between metal ions and peptides. These tools make it possible to virtually test

conformation and **metal affinity** hypotheses before proceeding to synthesis and experimental testing.

- **In vitro experiments**: Experimental tests are essential to validate theoretical predictions and confirm the efficacy of metal-peptide complexes in **biological media**. Experimental approaches such as **nuclear magnetic resonance (NMR) spectroscopy**, **UV-Vis spectroscopy**, or **fluorescence** can be used to analyze metal-peptide binding properties. In addition, **cytotoxicity** and **cell penetration assays** can be used to assess the efficacy and safety of complexes in real biological systems.

The integration of these different approaches makes it possible to design metal-peptide complexes with **enhanced predictability** in terms of stability, selectivity and biological activity. These advances are essential for the development of **targeted therapies** and **drugs** based on metal-peptide complexes, and for a better understanding of **metal-peptide interactions** in complex biological systems.

Chapter 9: New Perspectives in Metal Peptide Research

Metal peptides are at the forefront of many scientific innovations, with growing potential in fields as diverse as **nanotechnology**, **nanomedicine**, and **new genetic sequencing technologies**. This chapter explores **emerging perspectives** in metal peptide research and highlights their growing role in **medicine**, **protein engineering**, and **tomorrow's biomedical applications**.

9.1 Peptides and metals in tomorrow's biological applications: The emerging role of peptides in nanotechnology and nanomedicine

Nanotechnology and **nanomedicine** represent areas where the use of metal peptides could revolutionize the way medical treatments are delivered, particularly in **targeted therapies** and **diagnostics**. **Metal peptides** are increasingly used in the **manufacture of nanoparticles** and other nanoscale devices, thanks to their ability to bind specifically to metal ions and their unique bioactive properties. These peptides can be used to **build nanoscale structures** capable of delivering drugs or **targeting specific biological sites** with unprecedented precision.

For example, metal-peptide complexes can be used in the **preparation of metal nanoparticles** that serve as vectors for drugs or therapeutic genes. These nanoparticles, functionalized with metal peptides, can be programmed to bind to specific cellular receptors, facilitating **targeted delivery of** treatments directly into diseased cells, while minimizing systemic side effects.

Copper-peptide or **silver-peptide complexes**, for example, show great promise in the field of **antimicrobial nanoparticles**. They can be designed to bind to bacteria or tumor cells and release cytotoxic metal ions, paving the way for next-generation **antimicrobial** and anticancer **therapies**. The **antimicrobial** properties of metal-peptide complexes, combined with their ability to bind and deliver metal ions in biological environments, make these systems a promising avenue for combating **bacterial resistance** and **cancer**.

In addition, **nanomedicine** using metal peptides could enable the creation of **intelligent systems** that respond to environmental changes, such as variations in pH, temperature or other biological signals. These **nanomedicines** could release their active charge only under specific conditions, enhancing their therapeutic efficacy and safety.

9.2 The role of metal peptides in new genetic sequencing technologies and in protein engineering for new therapeutic applications

Another area where metal peptides are set to play a key role is in **new technologies for genetic sequencing** and **protein engineering**. **Metal peptides** can be used to develop powerful **biomolecular tools** for DNA **sequencing** and analysis. For example, **metal-peptide complexes** can be integrated into sensitive detection systems that exploit interactions between specific peptides and genetic targets to improve the efficiency and accuracy of DNA sequencing.

A prominent example of this use is the integration of **metal peptides** into **sequencing nanoparticles** to capture and analyze specific genetic sequences. Depending on the metal ions present and the peptide structure, these nanoparticles can be used for applications such as **third-generation sequencing** (e.g. nanopore sequencing), where metal peptides are responsible for recognizing certain nucleic bases and facilitating the passage of DNA molecules through nanopores for fast, accurate reading.

At the same time, metal peptides also play a crucial role in **protein engineering** for advanced therapeutic applications. **The modification of peptides to optimize their interaction with metals** enables the creation of new **fusion proteins**, which combine the properties of biologically active peptides and metal complexes. For example, metal peptides can be used to **stabilize therapeutic proteins** under extreme conditions, or to **target specific proteins** in anti-cancer treatments or gene therapies.

Therapeutic applications of metal peptides in protein engineering include their use in **gene therapy**, where metal peptides can be integrated into **viral or non-viral vectors** to enhance the **stability** and **selectivity of gene transport**. These metal-peptide complexes thus enable better **management of genetic pathologies** through **targeted delivery of** reparative or regulatory genes to diseased cells.

Metal peptides also have potential in the creation of **artificial proteins** or **enzyme systems** that mimic the functions of natural proteins, but are optimized for therapeutic applications. For example, metal peptides could be used to create **DNA repair enzymes**, facilitating the **repair of genetic mutations** in treatments for hereditary diseases.

Conclusion

The role of metal peptides in **new technologies** is expanding rapidly. Their ability to interact with metals and **modulate** these interactions opens the way to revolutionary applications in **nanomedicine**, **genetic sequencing** and **protein engineering**. Metal peptides are powerful tools for developing **new therapeutic treatments**, **innovative diagnostic systems** and **intelligent nanoparticles**. As research progresses, these peptides promise to transform the landscape of **targeted therapies**, **DNA repair** and **precision medicine**.

Chapter 10: Tools and Techniques for Exploring Metal-Peptide Interactions

In the field of peptide chemistry and metal-peptide complexes, a combination of analytical and spectroscopic techniques is essential to explore the interactions between metal ions and peptides. These interactions directly influence the structure, dynamics and biological function of the complexes, with applications in biotechnology, medicine and catalysis. Here's a detailed overview of the techniques most commonly used to study these complex systems:

1. Spectroscopy

UV-Vis (Ultraviolet-Visible)

UV-Vis spectroscopy makes it possible to examine electronic transitions in molecules, which is particularly useful for studying metal-peptide complexes. Metal ions often have specific absorption bands in the UV-Vis, which can be modified by coordination with a peptide. This technique enables us to monitor metal-peptide interactions as a function of metal concentration, complex geometry and chemical environment. It is used to characterize metal-peptide complexes, determine complex stoichiometry, and study changes in metal oxidation states.

Infrared spectroscopy (FT-IR)

FT-IR spectroscopy can be used to analyze the molecular vibrations of peptide functional groups, particularly those involved in coordination with metals. The characteristic vibrational bands of amide groups and amino acid side chains are influenced by the binding of the metal ion, enabling us to understand how the metal modifies the secondary and tertiary structure of the peptide. This technique is particularly useful for studying the structure and conformational changes induced by metal ion coordination, especially for peptides containing thiol or carboxyl groups.

Circular Dichroism (CD)

CD spectroscopy is a powerful technique for studying the secondary structure of peptides. It measures the differential absorption of circularly polarized light, making it possible to determine the conformation of peptides in solution. When metals are introduced, changes in the CD spectrum can reveal modifications to the secondary structure, such as changes in the proportion of α-helical or β-leaflet structures. This technique is crucial for observing the effect of metals on peptide conformation, particularly with regard to the induction of new structures or the stabilization of existing ones.

Fluorescence

Fluorescence is used to analyze the local environment around fluorophores in peptides, often synthetically introduced for spectroscopic studies. Changes in fluorescence intensity, emission peak position or fluorescence lifetime can be used to study complex interactions between metal ions and specific peptide residues. Fluorescence can also be used to study the kinetics of metal-peptide interactions in real time, which is particularly useful for complex binding and dissociation studies.

2. Nuclear Magnetic Resonance (NMR)

High Resolution NMR

NMR is the method of choice for determining the three-dimensional structure of peptides and metal-peptide complexes. It resolves structural details in solution, and can provide information on metal-peptide coordination interactions, side-chain flexibility and complex dynamics. NMR resolution is sufficient to obtain atomic information on metal-peptide coordination, molecular dynamics, and the formation of secondary and tertiary structures. This technique is therefore essential for understanding how metals influence peptide structure and function.

Paramagnetic NMR

Paramagnetic NMR is particularly useful for studying metal complexes containing paramagnetic ions, such as Fe^{3+}, Cu^{2+} and Zn^{2+}. These ions possess unpaired electron spins, which induce additional effects in NMR spectra that can be used to study metal-peptide interactions. Paramagnetic NMR enables the detection of metal-peptide interactions that are not visible with conventional NMR, thanks to amplified relaxation effects that are sensitive to metal proximity. This technique is useful for studying high-affinity metal-peptide complexes and specific interactions between peptide residues and the metal.

3. Chromatography

HPLC (High-Performance Liquid Chromatography)

HPLC is a key technique for purifying peptides and metal-peptide complexes. It enables peptides to be separated according to their chemical properties, and analyzed for their interaction with metals. It is often used to isolate metal-peptide complexes from complex mixtures, enabling pure samples to be obtained for further analysis. HPLC can also be used to quantify the concentration of metal-peptide complexes in solution.

Ion exchange chromatography

This method is used to isolate and analyze metal complexes from peptides. It relies on ion exchange between the metal complexes and an ion exchange material, enabling the complexes to be separated according to the nature and charge of the metal ions. It is particularly useful for studying metal transition complexes, and for purifying and separating metal-peptide complexes according to their respective affinities for a given metal ligand.

4. X-ray crystallography

X-ray crystallography enables us to resolve the three-dimensional structure of peptides and metal-peptide complexes at atomic resolution. The technique is based on X-ray diffraction of crystals of the molecule of interest. It enables us to visualize the exact arrangement of atoms in complexes, and to determine the coordination geometry between metals and peptides. X-ray crystallography is ideal for obtaining detailed information on the structure of metal-peptide complexes in crystals, and can be used to study complex interactions at the atomic level.

5. Mass spectrometry (MS)

Mass Spectrometry (MS)

Mass spectrometry is a powerful technique for determining the molecular mass of peptides and metal-peptide complexes. It can be used to analyze modified peptides, study metal complexes and determine peptide cleavage products for sequence identification. Tandem mass spectrometry (MS/MS) is often used to identify specific interactions between metal ions and peptide residues. This technique is also very useful for studying post-translational modifications of peptides, such as phosphorylation or glycosylation, in response to interaction with metals.

6. Microscopy

Transmission Electron Microscopy (TEM) and Scanning Electron Microscopy (SEM)

Electron microscopy is used to examine the structure and assembly of metal-peptide complexes at the nanoscale. TEM enables the visualization of individual complexes at very fine resolutions, while SEM provides surface images of complexes and aggregates on a larger scale. These techniques are particularly useful for examining complex formation and organization in biological materials or nanostructures.

Atomic Force Microscopy (AFM)

AFM is used to analyze the topography of peptides and metal-peptide complexes at the nanometric scale. It enables the study of interactions between peptides and surfaces, such as those of cell membranes, and can be used to observe structural modifications due to metal binding. The technique can also be used to measure the binding forces between peptides and metals.

7. Calorimetric analysis

ITC (Isothermal Titration Calorimetry)

ITC enables us to study the thermodynamic interactions between a peptide and a metal by measuring the variations in heat released or absorbed during the interaction. It provides valuable information on the affinity, stoichiometry and binding enthalpies of metal-peptide complexes, which is crucial for understanding the nature of the interactions.

DSC (Differential Scanning Calorimetry)

DSC can be used to analyze changes in the thermal stability of peptides and metal-peptide complexes. It measures the heat required to raise the temperature of a sample, providing information on the stability of complexes under different thermal conditions.

8. Molecular Modeling and Simulation

Numerical simulations, such as molecular dynamics and docking modeling, are used to predict and visualize interactions between peptides and metal ions. These tools enable us to simulate the behavior of metal-peptide complexes at the atomic scale, explore possible conformations and predict the dynamics of metal-peptide bonds as a function of various parameters, such as metal concentration and environmental conditions.

9. Electrochemistry

Electrochemistry is used to study the redox properties of metal-peptide complexes. Techniques such as cyclic voltammetry and polarography are used to analyze the oxidation state of metal ions in complexes and determine their affinity for peptides. These methods provide a better understanding of the stability of metal-peptide complexes as a function of the redox environment.

10. Molecular Biology and Functional Testing

Biological tests such as binding assays (Western Blot, ELISA) and cell-based assays are used to assess the biological effects of metal-peptide complexes. These techniques enable us to understand the impact of these complexes on proteins or biological receptors, and to test their effects in in vitro biological models.

Chapter 11: P2 and P3 laboratories in metal peptide research

A bioinorganic laboratory specializing in the chemistry of peptides and metal-peptide complexes, even when it includes advanced techniques such as PCR, sequencing, or other biological analyses, may be classified as P2 (biosafety level 2) or P3 (biosafety level 3), depending on the nature of the biological materials used and the potential risks associated with the work.

1 Biological safety level 2 (P2)

A level P2 laboratory is suitable for work with biological agents presenting a moderate risk to human health and the environment. This safety level is used for agents that are unlikely to cause serious illness in healthy workers, and for experiments that do not present a direct risk to the community or the environment.

Conditions for a P2 laboratory :

- **Biological agents**: P2 laboratories are generally used for biological agents such as non-pathogenic or low pathogenic bacteria, viruses or fungi. For example, bacteria such as non-pathogenic E. coli or laboratory strains of yeast.

- **Handling biological material**: The laboratory can handle DNA, RNA, human, animal or plant cell cultures, and model systems such as mammalian cells. This includes techniques such as PCR, cell culture or gene and protein analysis.
- **Safety**: Access to the laboratory is controlled, with personal protective equipment (PPE) such as gowns, gloves and eye protection. The use of biological safety hoods is required for certain procedures, particularly those generating aerosols (such as PCR).
- **Examples of work**: Within this framework, a bioinorganic chemistry laboratory can carry out experiments on peptides, metal-peptide complexes, and the analysis of gene expression or metal-biological interactions.

***Characteristics of a P2 laboratory* :**

- Installation: Class II biological safety hood, locking windows and doors, labelling of areas and reagents.
- Cleaning and decontamination protocol: Strict disinfection and sterilization protocols to eliminate all potentially hazardous biological materials.

2 Biological safety level 3 (P3)

A P3 laboratory is designed to handle biological agents which can cause serious illness in humans or animals, but which can be controlled by appropriate safety measures. This level of safety is stricter and is often used for pathogens, such as viruses, bacteria or other dangerous microorganisms that require extra precautions.

Conditions for a P3 laboratory **:**

- **Biological agents**: P3 laboratories are used for agents that can cause serious illness, such as pathogenic viruses (like SARS-CoV-2, tuberculosis, yellow fever virus, prions, etc.) or pathogenic bacteria with aerosol modes of transmission.
- **Safety**: Access is stricter, with HEPA air filtration systems, more rigorous personal protective equipment (such as hermetically sealed coveralls), and airflow control to avoid cross-contamination.
- **Examples of work**: P3 laboratories are more commonly used for biomedical research involving dangerous pathogens or viruses, but if you handle particularly pathogenic strains of viruses or biological agents as part of your research (for example, studying the effects of metals on viruses), a P3 laboratory may be required.

Characteristics of a P3 laboratory :

- **Enhanced safety**: Installation of negative pressure chambers, HEPA filtration systems for air outlets, safety devices for entry and exit (e.g. decontamination airlocks).
- **Protective equipment**: Use of protective suits, respiratory masks and other safety equipment when handling samples or cultures of hazardous microorganisms.
- **Strict protocols**: Manipulations are carried out in negative-pressure rooms and under a class III biological safety hood (where researchers work in sealed gloves and under additional protection).

4. **Application in a bioinorganics laboratory** :

   Laboratory P2

A bioinorganics laboratory studying interactions between peptides and metals, with the aim of understanding the chemistry of metal-peptide complexes or testing biochemical reactions in biological systems, would generally be classified P2, as it does not handle high-risk pathogens or genetically modified organisms (GMOs).

Techniques such as PCR and sequencing can be used in a P2 laboratory if the biological materials (e.g. DNA, RNA or proteins) are not associated with dangerous pathogens. You could be handling normal cell cultures or peptides, performing bioactivity tests, genetic analyses, etc.

Laboratory P3 :

A P3 laboratory may be required if you handle pathogens as part of your research, which is relatively rare in a bioinorganic chemistry laboratory. For example, if your research involves pathogenic viral strains or experiments on animal models infected with infectious agents or transmissible viruses, P3 classification would be required. P3 laboratories are mainly used in areas such as virology or microbiology studies with pathogens.

Other specific requirements :

A bioinorganic laboratory working on peptide and metal chemistry with cells or biological material might require additional bio-safety equipment to work with cell cultures, microorganism cultures or biological model systems. However, the nature of peptide chemistry and metal-peptide complexes does not normally require a P3 laboratory, unless pathogenic viruses or infectious agents are involved in the study.

Conclusion:

P2 classification would generally be sufficient for a bioinorganic laboratory focused on peptide chemistry and metal-peptide interactions, especially if you're working with human or animal cells, peptides, RNA or DNA, and techniques such as PCR and sequencing.

If biological pathogens are involved, or if viruses or other infectious agents are handled in your experiments (which is unlikely in a classical bioinorganic setting), then a P3 laboratory would be required.

The level of safety will depend primarily on the nature of the biological materials handled and the biological risks associated with your specific research.

Chapter 12: Case studies of recent discoveries

Metal-peptide complexes have emerged as powerful tools in the treatment of a variety of pathologies, from **neurodegenerative diseases** to **cancer** and **DNA repair**. This chapter presents recent case studies that illustrate how these complexes can be applied in innovative therapeutic contexts. Through these examples, we will demonstrate how metal peptides modulate key biological processes to improve existing medical treatments or create new ones.

12.1 Example 1: Cu/Zn complexes in the treatment of neurodegenerative diseases (Parkinson's, Alzheimer's)

Neurodegenerative diseases such as **Parkinson's** and **Alzheimer's** are linked to an abnormal accumulation of misfolded proteins, generating **oxidative stress** in the brain. One recent approach to treating these diseases involves the use of **metal complexes**, notably **Cu/Zn** complexes, for their ability to interact with proteins, stabilize their conformation and reduce oxidation.

The role of Cu^{2+} and Zn^{2+} ions in neurodegenerative diseases: Metal ions such as copper and zinc play an essential role in the **function of certain enzymes** and in the management of oxidative stress. For example, in **Parkinson's disease**, studies have shown that **Cu/Zn** complexes can interact with proteins such as **Parkin**, a protein involved in the degradation of misfolded proteins. These metal complexes can restore enzymatic activity and reduce the accumulation of toxic aggregated proteins, helping to slow disease progression.

Application example:

A study has shown that well-designed **Cu/Zn** complexes can be used to **stabilize neuroprotective proteins**, such as **superoxide dismutase** (SOD1), which is an enzyme involved in the detoxification of free radicals in nerve cells. These complexes have shown a promising ability to inhibit neurotoxicity in models of Parkinson's and Alzheimer's disease by modifying the redox properties of metals and facilitating **the reduction of protein aggregates**.

Outlook:

Research into the use of Cu/Zn complexes in the treatment of neurodegenerative diseases continues, with clinical studies in development to assess their long-term efficacy and safety.

12.2 Example 2: Platinum-based peptides for chemosensitization of cancer cells

Cancer treatments such as **chemotherapy** present considerable challenges, including **drug resistance** and toxic side effects. Metal peptides, particularly those containing **platinum complexes**, have shown great potential to overcome these challenges, especially as **chemosensitizers** to improve the efficacy of chemotherapies.

The role of platinum in chemosensitization :

Platinum, as a chemical element, is a key agent in many anti-cancer treatments, notably **cisplatin**. However, resistance to cisplatin remains a major obstacle. The use of **platinum-based metal peptides** partially overcomes this resistance, by enhancing **intracellular** platinum **accumulation** and altering platinum-DNA interaction, making cancer cells more sensitive to treatment.

Application example:

Researchers have developed **platinum-peptide complexes** that bind specifically to cancer cells via **cell surface receptors** and release platinum directly into the cell. These complexes showed a **significant increase in cytotoxic efficacy** against cisplatin-resistant cancer cells. What's more, the peptides used in these complexes can also **inhibit the**

DNA repair mechanisms of cancer cells, preventing their survival after treatment.

Outlook:

Platinum-based metal peptides continue to be explored for their potential to improve the efficacy of anti-cancer treatments, with ongoing studies into their toxicity, clinical efficacy, and ability to specifically target tumor cells without affecting healthy cells.

12.3 Example 3: Metal Peptides in the Activation of DNA Repair after Radiation Exposure

Activation of DNA repair is essential after exposure to **ionizing radiation**, whether during radiotherapy treatment or accidental exposure. **Metal peptides** can play a key role in activating **DNA repair** mechanisms by using metal complexes capable of interacting with repair enzymes.

The role of metal peptides in DNA repair :

After exposure to radiation, **DNA damage** such as double-strand breaks can occur, which is extremely toxic to the cell. **Metal-peptide complexes**, particularly those containing ions such as **zinc** or **copper**, can **activate enzymes** such as **parp (poly(ADP-ribose) polymerase)** or **DNA ligase**, essential in DNA repair processes. These complexes can stimulate enzymatic activity, enabling faster, more effective repair of **DNA breaks**.

Application example

A study has shown that **zinc-peptide** complexes increase the activity of **DNA ligase**, an enzyme that helps repair DNA breaks after exposure to radiation. These complexes showed a **protective effect** on irradiated human cells, reducing DNA damage and improving cell viability. In addition, these metal peptides can **stimulate repair of healthy cells** while minimizing damage to surrounding tissue.

Outlook:

These discoveries open up new avenues for the use of metal peptides in **radiation protection** and **DNA repair** after radiotherapy, a key area for the prevention of side effects associated with cancer and radiotherapy.

Conclusion

Metal-peptide complexes have shown remarkable therapeutic potential in a variety of medical fields, from **reducing oxidative stress** in neurodegenerative diseases to improving the efficacy **of anti-cancer treatments** and **DNA repair** after radiation. These **recent discoveries** underline not only the growing role of metal peptides in biology, but also their potential as therapeutic tools in the future. Continued research in these fields should open up new avenues for more targeted and effective treatments, with fewer side effects.

Chapter 13: Towards a New Frontier: Computer Science and Artificial Intelligence in the Design of Metal-Peptide Complexes

Research into **metal-peptide** complexes has evolved rapidly thanks to the integration of new technologies, notably **computer science** and **artificial intelligence (AI)**. These modern tools are **revolutionizing the way metal-peptide complexes are designed**, analyzed and optimized for various biomedical applications. This chapter explores the impact of computer simulations, AI algorithms and bioinformatics tools in the **design of metal peptides**, highlighting how these technologies can **reduce development time**, **optimize interactions** and **predict the biological properties** of complexes.

13.1 Simulations of Metal-Peptide Complexes via AI: Accelerating the Design of New Metal Peptides for Medicine

Metal-peptide complexes offer promising therapeutic prospects, but their design is a major challenge due to the **complexity of metal interactions** and the **diversity of peptide structures**. Traditionally, the design of these complexes has relied on time-consuming and costly experimental approaches. However, with the advent **of artificial**

intelligence and **machine learning algorithms**, these processes can now be significantly optimized.

The role of AI in the design of metal-peptide complexes: AI makes it possible to **model complex interactions** between metal ions and peptides, taking into account various thermodynamic and kinetic factors. These **computer simulations** enable us to predict with great precision how a metal will bind to a specific peptide, and the impact of this interaction on the structure and stability of the complex. AI also makes it possible to optimize parameters such as **metal selectivity**, **complex stability** and **pharmacokinetic properties**, thereby reducing the number of experimental trials required.

Application examples:

1. **Prediction of metal coordination sites**: Machine learning algorithms can identify the peptide residues most likely to bind to a given metal ion, based on databases of existing metal peptide structures. For example, neural networks can learn from known structures of metal complexes to predict how new peptides will interact with metal ions such as **copper**, **zinc** or **iron**.
2. **Optimization of pharmacological properties**: AI is used to optimize the **pharmacological properties** of complexes, such as

their **affinity for biological receptors** or their ability to cross cell membranes. These simulations help fine-tune peptides to maximize treatment efficacy while minimizing side effects.

3. **3D complex modeling**: **Molecular docking** and **molecular dynamics** algorithms can be used to simulate peptide-metal interactions in a three-dimensional environment. These tools can predict **complex stability, conformation** and **interaction with biological targets**, facilitating the design of more targeted and effective metal peptides.

Advantages of integrating AI into the design of metal-peptide complexes:

- **Cost reduction**: By simulating metal interactions, AI reduces the number of costly laboratory experiments.
- **Time savings**: Simulations speed up the discovery process, by quickly predicting which metal-peptide complexes could have interesting biological effects.
- **Personalizing treatments**: AI can enable the design of tailor-made metal peptides, adapted to specific therapeutic needs, notably in cancer treatments or DNA repair.

13.2 Examples of Bioinformatics Tools Used to Predict Metal-Peptide Interaction

Bioinformatics tools have developed rapidly in recent years, and many of them are now being used to explore and predict interactions between metal ions and peptides. These tools are playing a key role in **revolutionizing biomedical research** and paving the way for **new metal therapies**.

1.molecular docking software

Molecular docking software, such as **AutoDock**, **DockingServer** or **GOLD**, is used to predict the **conformation of metal-peptide complexes**. These tools take into account all the **interaction forces** (hydrogen, ionic, Van der Waals) between the metal and the functional groups of the peptide, in order to simulate the most stable structure of the complex. These tools enable us to select the best candidates for further study in the laboratory.

Example of use: Researchers have used AutoDock **to simulate the interaction of copper** with antimicrobial peptides, and have discovered **preferential coordination sites** for enhanced antibacterial activity.

2. Molecular dynamics (MD) modeling

Molecular dynamics simulations, such as those carried out with **GROMACS**, **AMBER** or **CHARMM**, simulate the evolution of metal-peptide complexes over time. These tools track **atomic movements,** taking into account metal interactions, conformational fluctuations and energy changes, enabling us to study the stability of complexes under different biological conditions.

Example of use: A study using **GROMACS** simulated the interaction of a lacticin-derived peptide with Cu^{2+} ions, predicting how this complex could induce better penetration of bacterial membranes, thereby enhancing its antimicrobial efficacy.

3. Software for predicting peptide secondary structures

Tools such as **PSIPRED**, **JPred**, or **I-TASSER** are used to predict the secondary and tertiary structure of peptides, and thus assess how the peptide's structure affects its ability to bind a metal. These predictions help to design metal peptides with the optimum structure for **efficient metal coordination**.

Example of use: Tools like **I-TASSER** are used to predict the three-dimensional structure of new peptides and estimate their metal binding sites, facilitating the design of new metal complexes with specific biological properties.

4. Artificial intelligence tools for peptide sequence optimization

Machine learning algorithms can be used to generate and optimize peptide sequences that bind specifically to metals, while maintaining desired biological properties. These tools use **neural networks** and **databases** to learn patterns in existing metal peptides and predict effective peptide sequences.

Example of use: A team of researchers has used **neural networks** to generate **metal peptides** capable of binding specifically to metal ions while having a lower affinity for human receptors, thus minimizing toxicity while maximizing therapeutic efficacy.

13.3 Revolution in Biomedical Research

Bioinformatics tools and artificial intelligence are having a **transformative impact** on **biomedical research**, particularly in the design of metal-peptide complexes. They not only **speed up the discovery of new therapeutic candidates**, but also **reduce the number of unsuccessful clinical trials**, by improving the accuracy of predictions on the structure and activity of metal-peptide complexes.

The main benefits for biomedical research include :

- **Optimizing drug development processes**: AI streamlines the drug discovery process, reducing the need for initial experimental testing.
- **Targeting more specific therapies**: By combining modeling and AI, it becomes possible to design highly specific **metal peptides**, capable of binding to precise biological targets.
- **Predicting side-effects and toxicity**: Algorithms can help predict the toxicity of metal-peptide complexes before moving on to clinical trials, thus reducing the risk of side-effects.

Conclusion

The integration of **computer science** and **artificial intelligence** in the design of metal-peptide complexes marks a major step forward in biomedical innovation. These technologies enable new complexes to be designed more efficiently, with **greater precision** and **lower costs**. In the near future, AI could not only accelerate the discovery of new treatments for complex diseases, but also enable the **creation of more**

targeted and personalized therapies for patients suffering from a variety of conditions.

Conclusion

Over the last few decades, the **chemistry of metal-peptide complexes** has seen spectacular advances, supported by scientific discoveries, technological innovations and multidisciplinary approaches. These metal complexes, combining **peptides** and **essential metals**, represent a rapidly expanding field, with increasingly diverse and promising applications in the **biomedical sector**.

Summary of recent developments

Recent developments in the design and application of metal-peptide complexes have transformed our understanding of their biological and therapeutic role. Significant progress has been made in the **synthesis of metal peptides**, in particular with the use of modern approaches such as **computational modeling** and **artificial intelligence**. These technologies enable **streamlined design** of metal-peptide complexes, promoting specific and effective interactions with biological targets while minimizing side effects.

The biomedical applications of metal-peptide complexes have also broadened, including not only the treatment of **cancer**, **neurodegenerative diseases** and **antibiotic-resistant infections**, but also **DNA repair** and **cellular protection** against oxidative stress. The combined **antioxidant**, **antimicrobial** and **anticancer** properties of metal-peptide complexes make these molecules ideal candidates for future cutting-edge therapies.

The Growing Importance of Metal Peptides in Tomorrow's Medicine

Metal peptides are now at the heart of **tomorrow's medicine**, particularly in **nanomedicine**, **gene therapy** and **targeted treatments**. Their **scalable properties**, the possibility of **fine chemical modifications** and their ability to interact with a variety of **essential biological metals** open up new therapeutic avenues. Integrating these complexes into **nanoparticles** or **nanomedicines** could improve the **targetability** and **selectivity of treatments**, offering more effective and personalized solutions for complex diseases such as **cancer** or **neurodegenerative disorders**.

What's more, **the synergy between peptides and metals** in **modular complexes** makes it possible to design treatments capable of overcoming **problems of drug resistance**, particularly in the context of **multi-drug-resistant infections** and **chronic diseases**. Metal peptides could also play a crucial role in **tissue repair** and **DNA repair**, particularly in targeted gene therapies.

Reflecting on the Future Contribution of Metal-Peptide Complexes in Innovative Biomedical Applications

As **basic research** and **technological advances** continue to advance, the future of **metal-peptide complexes** looks particularly promising. The use of **bioinformatics tools**, **artificial intelligence** and **nanotechnology** could lead to the **design of new classes of therapeutics**, offering more precise, less invasive and more effective solutions for a variety of medical conditions.

The applications of metal-peptide complexes are likely to extend beyond conventional treatments to include **therapeutic innovations** such as :

- **Targeted gene** and **drug delivery** via metal-peptide complexes encapsulated **in nanoparticles**, aimed at **increasing treatment precision** while reducing side effects.
- **Antiviral** and **antibacterial therapy**, where metal-peptide complexes will play a key role in **combating multi-resistant infections** and the emergence of new antibiotic **resistances**.

- **Tissue repair** and **cell regeneration**, where metals like zinc and copper, linked to specific peptides, could promote **DNA repair** mechanisms and **reduce oxidative stress**, opening up new horizons in the treatment of degenerative diseases and cellular aging.

In short, metal-peptide complexes represent an **exciting frontier** in biomedical research, with **revolutionary** potential for numerous therapeutic applications. The future of this field looks rich in possibilities, with continued advances in the **understanding of bio-inorganic interactions** and the **creation of new biomedical materials**, enabling improved **diagnosis**, **treatment** and **prevention** of human disease.

References:

1. Liu, Y., et al. (2023). "Artificial Intelligence in Drug Design: The Role of Machine Learning in Peptide-Based Metal Complexes." Frontiers in Pharmacology, 14, 334-345.
2. Singh, R., et al. (2022). "Predicting Metal-Ion Binding in Peptides Using Computational Approaches." Journal of Computational Chemistry, 43(8), 920-930.
3. Chen, J., et al. (2021). "AI and Bioinformatics in the Design of Metal-Based Therapeutic Peptides." Bioinformatics, 37(11), 1450-1460.
4. Yin, H., et al. (2020). "Peptides and Metal Complexes: The Future of Antimicrobial and Anticancer Therapy." Materials Science & Engineering C, 110, 110758.
5. Wang, Y., & Chan, W. C. (2006). The chemistry of peptide synthesis. In *Peptide Synthesis and Applications* (pp. 45-67). Wiley-VCH.
6. Gellman, S. H., & Angell, Y. L. (2007). The synthesis of peptides. In *Modern Peptide Chemistry* (pp. 211-256). Springer.
7. Henninot, A., & Le Couvey, P. (2019). Advances in the synthesis of peptides and proteins. In *Bio.*

8. Maret, W., & Li, Y. (2009). "The bioinorganic chemistry of zinc and its role in cellular processes." Biochemical Journal, 385(2), 1-18.
9. Bezerra, C. S., & Sette, L. D. (2020). "Copper in biochemistry and its role in enzyme catalysis." Biochemical Pharmacology, 178, 114119.
10. Gaggelli, E., et al. (2006). "The role of copper in biology and the role of copper in Alzheimer's disease." Bioinorganic Chemistry, 41, 11-17.
11. Puzon-McLaughlin, W., & Holmes, L. S. (2007).
12. Lippard, S. J., & Berg, J. M. (1994). "Principles of Bioinorganic Chemistry." University Science Books.
13. Crichton, R. R., & Wilmet, S. (2013). "Biological Inorganic Chemistry: An Overview." In *Biochemistry of the Transition Elements* (pp. 1-24). Springer.
14. Seybold, C., & Mertens, P. (2017). "Peptides as Metal-Binding Ligands in Medicine." European Journal of Medicinal Chemistry, 137, 269-283.
15. Meunier, B. (2016). "Metals in Bioinorganic Chemistry: An Introduction." Coordination Chemistry Reviews, 309, 1-10.

16. Pace, H., & Wenzel, T. (2019). "Metal-Peptide Complexes in Catalysis and Drug Delivery." Chemical Society Reviews, 48, 3516-3528.
17. Callewaert, C., et al. (2019). "Copper-Based Peptide Complexes for the Treatment of Multidrug-Resistant Bacterial Infections." Frontiers in Microbiology, 10, 2560.
18. Ebrahimian, M., et al. (2018). "Polymyxin B-Copper Complexes: A New Approach for Fighting Resistant Bacteria." Bioorganic & Medicinal Chemistry Letters, 28(7), 1152-1156.
19. Ghosh, S., & Ghosh, R. (2017). "Peptides Derived from Lacticin for Their Antimicrobial Activities and Copper Complexation." Journal of Applied Microbiology, 122(3), 577-588.
20. Sarma, B. K., et al. (2020). "Antimicrobial Peptides: A New Horizon in the Treatment of Drug-Resistant Infections." International Journal of Molecular Sciences, 21(11), 3775.
21. Jia, X., et al. (2018). "Platinum-based Peptide Complexes for Targeted Cancer Therapy: Mechanisms and Applications." Journal of Medicinal Chemistry, 61(12), 4779-4792.
22. Jiang, X., et al. (2020). "Peptide-Metal Nanoparticles as Vectors for Gene Delivery in Cancer Therapy." Biomaterials, 265, 120393.

23. Frye, J. S., et al. (2019). "Therapeutic Peptides in Cancer Treatment: Current Challenges and Future Directions." Future Medicinal Chemistry, 11(4), 433-448.

24. Bertini, I., Luchinat, C., & Rizzotti, S. (2001). "NMR of metalloproteins and metalloenzymes." In *Bioinorganic Chemistry: Inorganic Elements in the Chemistry of Life* (pp. 345-378). Wiley-VCH.

25. Berkowitz, A. L., et al. (2017). "Peptide-Metal Complexes in Antioxidant Defense: Mechanisms and Applications." Journal of Inorganic Biochemistry, 173, 75-88.

26. Lee, S. H., et al. (2018). "Glutathione and Manganese Complexes in Cellular Defense Against Oxidative Stress." Free Radical Biology and Medicine, 120, 324-331.

27. Gomez, D. A., et al. (2019). "The Role of Cu/Zn-Superoxide Dismutase in DNA Repair and Cellular Protection Against Oxidative Damage." Cellular and Molecular Life Sciences, 76(2), 261-276.

28. Caldwell, S. R., et al. (2019). "Challenges in the Design of Metal-Peptide Complexes for Biomedical Applications." Chemical Reviews, 119(10), 5276-5297.

29. Zhang, Q., et al. (2020). "Peptide-Based Metal Complexes: Rational Design and Application in Drug Delivery." Biomaterials, 231, 119681.
30. Liu, X., et al. (2018). "Designing Metal-Peptide Complexes for Targeted Therapy: A Computational Approach." Journal of Medicinal Chemistry, 61(18), 8407-8416.
31. Lee, H., et al. (2020). "Peptide-Based Nanoparticles for Targeted Drug Delivery and Therapeutic Applications." Journal of Controlled Release, 320, 176-187.
32. Koch, C., et al. (2019). "Metal-Peptide Complexes in Nanomedicine and Their Biomedical Applications." Biomaterials, 211, 61-74.
33. Stolz, F., et al. (2021). "Peptide Metalloproteins in Biotechnology and Medicine: Engineering, Applications and Future Perspectives." Nature Communications, 12, 6251.
34. Banci, L. (2014). *Handbook on Metalloproteins*. Wiley-VCH.
35. Sigel, A., & Sigel, H. (2004). *Metallomics: The Science of Metals in the Biological World*. Springer.
36. Lakowicz, J. R. (2006). *Principles of Fluorescence Spectroscopy* (3rd ed.). Springer.

37. Cohen, S. M., & Dempsey, J. L. (2015). *Coordination Chemistry of Metal-Peptide Complexes*. Wiley-VCH.
38. Chen, W., & Brooks, C. L. (2010). *Computational Methods in Protein Structure and Dynamics*. Springer.
39. Snyder, R. G., & Cava, R. J. (1999). *X-ray Crystallography: A Handbook for Structural Chemists*. Wiley.
40. Lennox, R. W., & Langford, J. K. (2013). *Introduction to Bioinorganic Chemistry*. Oxford University Press.
41. White, C.W., & White, D.C. (2004). *Spectroscopic and Electrochemical Methods in Inorganic Chemistry*. Elsevier.
42. Lippard, S. J., & Berg, J. M. (1994). *Principles of Bioinorganic Chemistry*. Mill Valley, CA: University Science Books.
43. Harrison, M. D., & Bell, R. M. (2000). *Peptides and Peptidomimetics as Drugs*. Springer.
44. Saito, N., & Kumagai, T. (2003). *Bioinorganic Chemistry of Metal Complexes with Biomolecules*. Wiley-VCH.
45. Bertini, I., Luchinat, C., & Neri, S. (2010). *Biomolecular Chemistry of Peptides and Metalloproteins*. Oxford University Press.
46. Dixon, D. P., & Edwards, R. (2013). *Antioxidants and the Role of Metal Complexes in Oxidative Stress*. Springer.

47.El-Batran, S. G., & El-Sayed, A. M. (2014). *Peptide-based Biomaterials: Synthesis and Applications*. Elsevier.

48.Rosenberg, B., & Van Camp, L. (2003). *Platinum-based Drugs in Cancer Therapy*. In *Cancer Chemotherapy: A Guide for the Practicing Oncologist*, 2nd Edition. Humana Press

Glossary:

Amino acids: Organic compounds that bind together to form peptides and proteins.

Peptide bond: Covalent bond formed between the amine group of an amino acid and the carboxyl group of another amino acid.

Resin synthesis: A method of peptide synthesis in which amino acids are added sequentially to a solid resin.

Recombinant peptides: Peptides produced by host cells after expression of the gene coding for the peptide of interest.

Therapeutic peptides: Small chains of amino acids used in the treatment of various pathologies, including cancer and genetic diseases.

Antimicrobial peptides (AMPs): Small chains of amino acids with the ability to kill or inhibit the growth of microorganisms, including bacteria, fungi and viruses.

Peptidomimetics: Chemical compounds mimicking the structure of peptides, but modified to improve their stability or affinity for certain receptors.

Metallomimetic peptides: Peptides designed to mimic the properties of a specific metal or metal complex.

Metal peptides: Peptides that form complexes with metal ions, modifying their biological properties.

Metallothionein: Protein that binds heavy metals, playing a role in metal regulation and detoxification.

Zincoenzymes: Enzymes containing zinc in their structure, crucial for their catalytic activity.

Ferritin: Protein that stores iron in cells.

Metalloproteins: Proteins containing a metal within their structure, essential for their catalytic activity.

Metal complexes: Assemblies formed by a central metal bound to ligands, often in proteins or enzymes.

Metal-peptide complexes: compounds formed by the coordination of a metal ion (such as Cu^{2+}, Zn^{2+}, Fe^{2+}) to a peptide. These complexes can modulate the peptide's biological activity.

Monodentate complex: Complex in which a single functional group binds to a metal.

Bidentate complex: Complex in which two functional groups bind to a metal.

Polydentate complex: Complex in which several functional groups bind to a metal.

Coordination: Chemical interaction between a metal and ligands, often via sulfur, nitrogen or oxygen atoms.

Chelation: Process by which a metal forms stable bonds with several atoms of a ligand (such as a peptide).

Dynamic chelation: Process by which a ligand, often a peptide, reversibly binds to and detaches from a metal ion, enabling fine regulation of metal-ligand interactions.

Coordination mechanism: The process by which a metal binds to a ligand, such as a peptide, to form a stable complex.

Bioinorganic catalysis: Use of metal complexes in chemical reactions promoted by catalytic sites, mimicking the action of natural enzymes.

Oxidative stress: Condition in which the production of free radicals exceeds the body's antioxidant capacity, leading to cellular damage.

Free radicals: Atoms or molecules possessing an unpaired electron, making them highly reactive and capable of causing biological damage, notably to DNA and cell membranes.

Superoxide dismutase (SOD): Antioxidant enzyme that catalyzes the conversion of superoxides to hydrogen peroxide, thereby helping to manage oxidative stress.

Cu/Zn-SOD: Copper- and zinc-containing form of superoxide dismutase, essential for defense against superoxide radicals.

Glutathionine: Antioxidant tripeptide involved in cellular protection against oxidative stress.

Bioreactor: Device in which biological processes take place under controlled conditions, used to grow cells or produce biomolecules such as peptides.

Gene therapy: Treatment consisting in inserting, deleting or modifying genes inside an individual's cells to treat genetic or acquired diseases.

Targeted therapy: Therapeutic approach involving the targeted delivery of drugs to a specific site in the body to reduce side effects and improve treatment efficacy.

Nanoparticles: Particles of nanometric size (generally between 1 and 100 nm) used in biological, chemical or industrial applications, often as vectors for the delivery of drugs or therapeutic genes.

Genetic vectors: Tools used to introduce therapeutic genes into cells, including plasmids, modified viruses or nanoparticles.

Chelator: Molecule capable of forming stable complexes with metal ions, often used to modify the selectivity of metal-peptide complexes.

Molecular dynamics: Numerical simulation of the movements of atoms in a molecule to predict its interactions and conformation.

Docking: Computer modeling method for predicting the interaction between a small molecule (such as a metal) and a macromolecule (such as a peptide).

Nanomedicine: Application of nanotechnology to treat diseases, notably by using nanoparticles to target and deliver drugs or therapeutic genes.

Genetic sequencing: Technique for determining the order of nucleotide bases (DNA or RNA) in a genome.

UV-Vis (Ultraviolet-Visible) spectroscopy: A spectroscopic technique that measures the absorption of light in the ultraviolet (UV) and visible (Vis) ranges by molecules. It can be used to study electronic transitions in molecules, such as metal-peptide complexes.

FT-IR (Fourier Transform Infrared) spectroscopy: Spectroscopic technique that measures the absorption of infrared radiation by molecules. It is used to examine the vibrations of functional groups in peptides, particularly when coordinated with metal ions.

Circular Dichroism (CD): Spectroscopy that measures the differential absorption of circularly polarized light, making it possible to study the secondary structure of peptides and conformational changes induced by metal binding.

Fluorescence: A method that measures the emission of light by a molecule after excitation by a light source. It is used to observe the local environment of fluorophores in peptides and metal-peptide complexes.

Nuclear Magnetic Resonance (NMR): A technique for studying the three-dimensional structure of peptides and metal-peptide complexes at the atomic scale. It is based on the interaction of atomic nuclei with an external magnetic field.

Paramagnetic NMR: A variant of NMR used for complexes containing paramagnetic metal ions, such as Fe^{3+} or Cu^{2+}. These ions have unpaired electron spins, which affect the NMR spectra.

HPLC (High Performance Liquid Chromatography): Technique for separating molecules in a sample according to their chemical properties, enabling peptides and metal-peptide complexes to be purified and analyzed.

Ion exchange chromatography: Technique for separating ions according to their charge and affinity for an ion exchange material, used to isolate metal-peptide complexes.

X-ray crystallography: X-ray diffraction technique used to determine the atomic structure of molecules in crystals. It can be used to visualize the geometry and interactions of metal-peptide complexes.

Mass Spectrometry (MS): Analytical technique used to measure the mass of molecules. It is used to determine the molecular composition of peptides and metal-peptide complexes, and to identify post-translational modifications.

MS/MS (Tandem Mass Spectrometry): A variant of mass spectrometry used to fragment ions and analyze the fragmentation products to determine the sequence of peptides and their metal interactions.

Printed by Books on Demand GmbH, Norderstedt / Germany